I0470299

PREPARATION AND DESCRIPTION OF SOLUTIONS

PREPARATION AND DESCRIPTION OF SOLUTIONS

In The Laboratory

Adolf Kofi Awua

Cellular and Clinical Research Centre,
Radiological and Medical Sciences Research Institute,
Ghana Atomic Energy Commission

iUniverse, Inc.
Bloomington

Preparation and Description of Solutions
In the laboratory

Copyright © 2011 by Adolf Kofi Awua

All rights reserved. No part of this book may be used or reproduced by any means, graphic, electronic, or mechanical, including photocopying, recording, taping or by any information storage retrieval system without the written permission of the publisher except in the case of brief quotations embodied in critical articles and reviews.

The views expressed in this work are solely those of the author and do not necessarily reflect the views of the publisher, and the publisher hereby disclaims any responsibility for them.

Inasmuch as the author believe that usage of equipment, devices, and chemicals, and other information as set forth in this book are in accord with current recommendations and standard practice at the time of publishing, he accept no legal responsibility for any errors or omissions and make no warranty, expressed or implied, with respect to material contained herein. The reader is encouraged to review and evaluate the information provided and instructions set out herein in view of ongoing research and the constant flow of information relating to laboratories for added warnings and precautions.

iUniverse books may be ordered through booksellers or by contacting:

iUniverse
1663 Liberty Drive
Bloomington, IN 47403
www.iuniverse.com
1-800-Authors (1-800-288-4677)

Because of the dynamic nature of the Internet, any Web addresses or links contained in this book may have changed since publication and may no longer be valid.

Any people depicted in stock imagery provided by Thinkstock are models, and such images are being used for illustrative purposes only.

Certain stock imagery © Thinkstock.

ISBN: 978-1-4502-7444-9 (sc)
ISBN: 978-1-4502-7445-6 (ebk)

Printed in the United States of America

iUniverse rev. date: 1/27/2011

To My Parents

PREFACE

The need for such a book became clear to me during the four years I worked as a Teaching and Research Assistant at the Biochemistry Department of the University of Ghana. It is difficulty to find a separate course dedicated to teaching students how to prepare solutions, particularly at the high school level; As a result, most undergraduate students find it difficult preparing solutions and graduate without an organized body of knowledge on how to prepare solutions. Also, the advent of kits, automated and semi automated procedures for most biological analyses have led to the lost of skills and know-how regarding the preparation of solutions by most laboratory personnel. Written in a very simple and reader-friendly manner, this book uses a stepwise approach in equipping readers with the relevant understanding and guidance that is needed in the preparation of solutions.

ACKNOWLEDGEMENT

The preparation of this book has been an effort based on my interactions with the lecturers and students I have worked with throughout my stay at the Biochemistry Department of the University of Ghana, particularly working with Dr. Yaa D. Osei and students of the molecular biology practical class.

I gratefully acknowledge the invaluable assistance of the Miss. Chantal A. S. Agbemebiese (then Graduate Student of the Biochemistry Department, University of Ghana) for reviewing, suggesting and providing the encouragement that pushed me to go beyond the initial manuscript. I also thank Miss. Francisca Emefa Kumadey for her review and suggestions. Also, I am grateful to Prof. N. A. Adamafio (University of Ghana) for all her assistance (both directly and indirectly), and to all who did or said one thing or the other to get this work done.

C O N T E N T S

SECTION

CONDUCT IN THE LABORATORY

Handling Chemicals

The laboratory has been of great value to human development and has mostly been the small room in which great discoveries have emanated. Although some discoveries in the laboratory were by serendipity, and sometimes as a result of non standard laboratory conduct (the discovery of antibiotics by Sir Alexander Fleming), the truth still stands that good conduct in the laboratory cannot be over looked.

Good laboratory conduct requires that all chemicals should be considered dangerous to human, animal and plant life and therefore, should be handled with extreme care. Depending on the type of laboratory, there are specific safety rules/guidelines to follow. However, this book presents some general safety tips that will be useful for preventing and/or minimizing exposure to harmful substances through inhalation, adsorption on skin and ingestion. When in doubt, the personnel in charge of the laboratory should be consulted. The safety tips are;

1. Protective wares such as lab coats, lab aprons, gas masks, gloves, goggles etc. are to be used when working in the laboratory.

2. All hot solutions should be handled with the necessary heat protective ware (gloves, tongues, racks and tube holders).

3. Chemicals are not to be touched directly, tasted, or smelled.

   If there is the need to smell a chemical, then gently fan the air above the opening of the chemical container toward the nose while breathing normally.

4. All organic solvents, fuming solvents and volatile solids should be handled in a fume hood or in other well ventilated place.

1

5. Mouth pipetting should be avoided at all times.

6. Solutions and other preparations should be properly/ appropriately collected and disposed off. Avoid damping all solutions directly down the drain through the sink. For example, solution containing microorganisms should be aseptically collected, autoclaved and burnt.

7. Before removing any chemical from its bottle or container, the user must double-check for the full information regarding the content and the necessary safety precautions. These necessary warnings are notably displayed on the labels of all chemicals that are hazardous. Also, most manufactures provide as part of their supplies, documentations that include an up-to-date Material Safety Data Sheet (MSDS). This document provides the user with the safety information needed to store, handle, transport and dispose these chemicals safely. For most manufactures, users can order free additional updates of the MSDS.

8. When working with toxic chemicals, the duration and frequency of exposure must be reduced.

9. When weighing or measuring out any chemical, taking smaller amounts at a time is often better than larger amounts since **unused chemicals should never be returned to the original container, be it liquid or solid.**

    However, when some amount of the chemical remains unused, it must be transferred into a new, well labelled air tight container and stored as the original chemical was stored.

10. Label all chemical containers immediately before transferring in the chemical.

11. All reagents and cleaned glassware should be returned after use to their proper place of storage.

12. Clean-up the working area before and after working in a laboratory, washing your hands thoroughly just before leaving the laboratory.

13. In case of an accident (slippage, cuts, etc.), immediately inform the personnel in charge.

Highly Toxic Chemicals

Although all chemicals should be treated as potentially hazardous, the chemicals listed in the table below have been shown to present a severe risk to human health.

It would be useful to note that safety measures regarding the handling of hazardous chemicals and working in the laboratory environment are not meant only for the safety of those working with these chemicals and in these laboratories but also for those who may either directly or indirectly come into contact with these persons. Therefore, **all protective wares should be left in the laboratory when moving out of the laboratory,** so as to avoid the possibility of transferring harmful materials from these protective wares to other persons. This list is not comprehensive but consists of some of the toxic chemicals in a laboratory.

Acetic anhydride	Acrylamide
Ammonia	Aniline
Bromine	Chlorine
Chromium compounds	Cyanides
Dimethyl sulfate	Ethanolamine
Formaldehyde	Formic acid
Glass wool fibres	Hydrochloric acid
Iodine	Hydrogen peroxide
Mercury and mercury compounds	Nitric acid
Nitrobenzene	N,N-dimethylaniline
Phenol	Phenyl hydrazine

Safety Symbols

The following safety signs may be displayed on the labels of chemicals and in the laboratory. It would be useful to look for more of these symbols and understand what they mean. The internet is a useful resource.

Biohazard

Combustible

Corrosive Material

Electrical Hazard

Environmental Hazard

Explosive Material

Eyewash

Flammable

General Hazard

Irritant

No Open Flames

Non Ionizing Radiation

Non-Potable Water

Radioactive

Toxic Chemical

SECTION

2

MEASUREMENTS AND FIGURES

Measurements

The preparation of solutions is preceded by a set of calculations and involves a series of measurements using instruments and apparatuses. These instruments are limited by the number of decimal places they can measure to. For example, some weighing balances can only measure masses to two decimal places (10.03 g), while others can measure to five decimal places (10.03543 g). The number of decimal places that can be obtained using an instrument is referred to as the sensitivity of the instrument. This should clearly be differentiated from the precision of an instrument, which is the ability of a particular instrument to reproduce the same measurement for a particular substance.

Although 100% precision is not attainable by any instrument, a very high degree of precision is desired. As an example, when a weighing balance was used to determine the mass of an empty beaker, the following were obtained on four occasions; 5.74**3** g, 5.74**2** g, 5.74**4** g and 5.74**3** g. These measurements vary with a small range of 0.002, indicating a high degree of precision. The question that needs to be answered here is, which of these masses is the correct measurement? Another situation is observed when reading the volume of a liquid in a measuring cylinder where the meniscus falls between two markings on the calibration. The last digit of this measurement will vary depending on the person reading the volume.

It therefore becomes clear that the last digit of a measurement cannot be reliable. With this understanding, it is obvious that when a particular measurement is sought, an instrument that can measure at least one decimal place longer than the desired decimal place is used. For instance, when measuring a weight of 4.2 g, a balance that can measure to two or more decimal places should be used and a measurement of 4.20 g should be aimed at. This concept of reliable digits in a measurement is what is referred to as significant figures.

Significant Figures

These are said to be the reliable figures in a measurement or the digits in a number known with high certainty. This means that all figures in a measurement should be noted. In doing this, zeros present a peculiar need. The digit "0" may have one of two roles in any number. In one situation, it is a number just like any other digit in the number. In the other situation, it is only an indicator of decimal places. For example, the number 2304 has four digits and all of these digits are important in stating the value of this number. However, for a number like, 0.05, the value does not depend on the zeros. Since it is a fraction where the zeros (0.0) only indicate the position of the decimal point.

With this peculiar nature of zeros, a set of rules are used to determine which digits are significant in a measurement.

1. Non-zero digits in a measurement are always significant.

   Example: *7437*   *: four significant figures*

                2.3452 : five significant figures

2. A zero or a series of zeros between any two non-zero digits are significant.

   Example: *203*   *: three significant figures*

             7.00081   *: six significant figures*

3. Any zero to the right of a decimal point in a measurement and at the end of the number is significant

   Example: *1.0*   *: two significant figures*

             1.0420 : five significant figures

4. Any zero before the first non-zero digit in a measurement is not significant. Remember that these zeros only indicate the position of the decimal point.

   Example: *0.0079 : two significant figures*

0.0498 : *three significant figures*

5. Zeros at the end of whole numbers may or may not be considered as significant.

    Example: 25000 : *two or five significant figures*

Scientific notations are used to specify the number of significant figures in whole numbers ending with zeros. To do this, a decimal point is placed between the first and second non-zero digits. The number is then written this way followed by the multiplication symbol and then by 10 raised to a power equal to number of digits after the decimal point.

Our example becomes: $2.\underline{5000} \times 10^4$

This number can then be written to any other number of significant figures as follows:

2.5000×10^4 : *five significant figures*

2.500×10^4 : *four significant figures*

2.50×10^4 : *three significant figures*

2.5×10^4 : *two significant figures*

After taking measurements and reporting them to desired significant figures, the numbers are used for calculating other parameters/properties of solutions, such as the concentration. Since these measurements may all not be reported to the same significant figure, the answer obtained after calculation may contain a different number of digits. The question to answer here is, to what number of significant figures should the answer be reported and how should the digits in the answer be reduced to that specific significant figure? The concept of rounding off will help in answering this question.

Rounding Off

This concept does not determine an accurate answer for a calculation; it is rather a convention of reporting an answer after a calculation.

This convention requires that only the final answer is to be rounded off in a series of calculations.

To round off a number to a particular significant figure the following rules may be used.

1. If the digit to be dropped is less than 5, then that digit and any other digit to its right are dropped.

   Example: Report the mass 2473871 µg to three significant figures.

   > *This implies that only three digits (from left to right) should be retained i.e.* **2473871** µg. *By this rule, since 3 is less than 5, the digit 3 and the digits after it should be dropped. The number then becomes* **247000** µg.

2. If the digit to be dropped is greater than 5, it is dropped and 1 is added to the digit immediately before it.

   Example: Report 15.3748 mL to three significant figures.

   > *Retaining three digits in the measurement* **15.3748** mL, *means that 7 and the numbers after it should be dropped. Since 7 is greater than 5, 1 should be added to the 3. Therefore, the number becomes* **15.4** mL.

3. If the digit to drop is 5 and is followed by a non-zero digit, add 1 to the digit immediately preceding the 5 and drop the 5 and any other digit after it.

   Example: Round off 16.6853 g to a four significant figure.

   > *Retaining four digits in the measurement* **16.685354** g *requires that 5 and any other digit after it be dropped. Since the digit after 5 is not zero, 1 should be added to 8. This number is then reported as* **16.69 g**.

4. If the digit to be dropped is 5 and it is followed by zero, and if the digit before the 5 is even, then drop the 5 and any other digit after it.

   Example: Round off 377.2650 g to five significant figures.

   > *Writing the* **377.2650** g *to contain five digits will demand that 5 and the zero after it be dropped. By this rule, this number becomes* **377.26 g,** *since the digit immediately before 5 which 2 is an even number.*

5. If the digit to be dropped is 5 and it is followed by zero, and if the digit before the 5 is odd, then add 1 to the odd digit and drop the 5 and any other digit after it.

    Example: Rewrite the number 21.79350 mg to contain five significant figures.

    > *To do this, the number* **21.793̲50** mg *should lose the digits* 5̲ *and 0. Since the digit before the* 5̲ *is odd, 1 is added to it to make the number* **21.794** mg.

Now that we understand how to round off to a particular number of significant figures, we will look at how to determine the number of significant figures to report for an answer after calculating for it. The rules vary for each type of mathematical operation.

Multiplication / Division

After multiplying or dividing two or more measurements, the answer should be reported to the least number of significant figures in the measurements.

Example: 0.046 cm × 374 cm × 2845 cm = 48945.38 cm^3

> *The measurement 0.046 cm has the least number of significant figures (two). Therefore, the answer should be reported to two significant figures. The answer becomes* **49000** *cm^3 or 4.9 × 10^4 cm^3.*

Example: 25.45 g ÷ 0.4 cm^3 = 63.625 g/cm^3

> *Since 0.4 cm^3 has the least number of significant figures (one), the answer becomes 60 g/cm^3.*

Addition / Subtraction

When measurements are to be added, or one is to be subtracted from another, the answer is recorded to the same number of decimal points as that of the measurement with the least number of **decimal places**.

Example: 434.372 g + 11.2 g = 445.572 g

> *The measurement 11.2 g has the least number of decimal places (one) therefore, the answer should be written as* **445.6** *g.*

These rules do not apply to numbers obtained by counting, since numbers obtained by counting are exact. For example, in calculating the average of four measurements of the volume of a reaction mix, the sum of the volumes is divided by four. The denominator, which is four, is not used to determine the number of significant figures to which the answer should be stated.

These rules are important for the determination of masses to weigh and volumes to measure when preparing simple solutions.

SECTION

3

SIMPLE SOLUTIONS

Description of Solutions

A **Solution** is defined as a combination of two or more substances such that individual components are uniformly distributed and not identifiable, in other words, a solution is a homogenous mixture of two or more substances.

This implies that for every portion of a solution, the relative amounts of the combining components would remain the same. These individual components of a solution are mostly referred to as follows:

A Solute is the component of a solution that dissolves or disperses when a solution is prepared. It is usually in smaller quantity. This can either be solid, liquid or gas.

A **Solvent** is the component into which the solute dissolves or disperses and the component usually in larger quantity. These are mostly liquids.

For example, adding a table spoon full of sugar to a cup of water makes a solution which has the sugar as the solute and water as the solvent. It should be noted that there can be more than one solute in a solution as is the case when sugar and salt are added to water to make a solution known as oral rehydration salt (ORS).

Solutions are qualitatively and quantitatively described based on the amount of solute they contain. The qualitative description of a solution involves the use of relative terms, such as less concentrated, more dilute etc. in stating the amount of solute it contains.

In describing a solution quantitatively, the exact amount or quantity of the solute is stated as a ratio of a unit or specific quantity of the solution or solvent. This is referred to as the **Concentration** of the

11

solution. When the number of mole(s) of the solute is stated as a ratio of a unit volume (measured in dm^3 or litre) of solution, the concentration is referred to as the **Molarity** (mol/dm^3 or mol/L).

On the other hand, when the mass of the solute is stated in relation to a unit volume (measured in dm^3 or litre) of solution, then the concentration is referred to as the **Mass Concentration** (g/dm^3, g/L, or mg/mL). *Note that this is the density of a solution that contains only one solute.*

When the concentration of a solution is expressed as the quantity of the solute to a 100 mL or 100 g of solution, the concentration is referred to as a **Percent concentration.** This term of concentration may be expressed as **Mass percent** (weight of solute/weight of solution), **Volume percent** (volume of solute/volume of solution) or **Mass-Volume percent** (mass of solute/volume of solution). As an example, if a 25 g of NaOH is added to 75 g of H_2O, a 100 g of solution is obtained that is 25% (w/w) NaOH. A 70% (v/v) isopropanol solution means that 70 mL of isopropanol was dissolved in enough water to obtain a 100 mL solution. Note that this may not be the same as adding 30 mL of water to the 70 mL of isopropanol. This is because, based on the partial molar properties of solution, the final volume of a solution is dependant on the mole of the substances in solution but not on the volumes of the solute and solvents.

A 5% (w/v) NaCl solution means that 5 g of NaCl is in a 100 mL of solution. You may try interpreting the concentration of a solution expressed as 23% (w/v).

Another term of concentration that can be stated either by volume or by mass of the solute is the **part per million** (ppm) terminology. This term is used when the solute is in very small amounts in a large amount of solution. For example: the volume of an agrochemical in water bodies (rivers and lakes) or the mass of pollutants in soil. Concentrations are stated as part per million by volume, to mean, the volume of the solute in micro litres per one litre of solution. When stated by mass, it is a measure of the mass of the solute in milligram per kilogram of solution.

Two concentration terms state the amount of solute relative to the quantity of solvent; **Solubility** – the concentration of the saturated solution of a solute, when this is measured in molarity it referred to as the molar solubility. **Molality** – the moles of solute per kilogram of solvent.

When a solution is a combination of more than two other solutes/solutions, the above descriptions of concentration will not be appropriate since the components of such a solution would include more than one solute. One of the quantitative descriptions of such solutions involves the use of **Ratio Content**. For example, a DNA extraction solution containing phenol: chloroform: isoamyl alcohol is described as a 25:24:1 solution. This means that 25 units of phenol are to be mixed with 24 units of chloroform and 1 unit of isoamyl alcohol to make this solution. This can therefore mean that when 25 mL, or any unit of measurement of phenol is used, 24 and 1 of the same unit should be used for the other components respectively.

Another mode of describing the concentration of a combination of solutions is the **Working Concentration** system. The unit of this description is X (times). This system is used when a solution is made-up of two or more different solutes. Each of these solutes may be at a different concentration. When the individual solutes are mixed so that each would be at the concentration required for use, then the solution is said to have a concentration of 1X.

For example, a 1X TAE solution contains 50 mM Tris-base, 10 mM acetate and 100 mM EDTA. Then a 10X TAE solution would contain 500 mM Tris-base, 100 mM acetate and 1 M EDTA. Each of the component solute are at 10X the working concentration. A 6X DNA loading solution contains 0.03% bromophenol blue, 0.03% xylene cyanol, 0.4% orange G, 10 mM Tris-HCl (pH 7,5) and 50 mM EDTA. What would be the concentration of the individual components for this DNA loading buffer that has a concentration of 1X?

Solutions can also be described according to their concentration; a solution of known concentration is referred to as a **Standard Solution**. A **Molar Solution** on the other hand is a solution that contains one mole of the solute in one litre of solution **(1 M)**. Such solutions contain a mass of the solute equal to its molecular weight in a litre of the solution. Understanding molar solutions makes it easier to prepare solutions of any concentration. For example, 1 M solution of sodium chloride of volume 1000 mL can be prepared by measuring 58.46 g of sodium chloride (the molecular weight of sodium chloride as it is on the label of its container), dissolve and make up the volume to 1000 mL (1 dm^3 or 1000 cm^3) with distilled water. In the same light, 1 M solution of sodium chloride of volume 100 mL will require 5.846 g of sodium chloride.

Clearly, the volume changes by the same factor as that of the mass. Therefore, by the use of simple proportion, it is possible to calculate the mass of a chemical substance to weigh in order to prepare any concentration and volume of its solution. Using the formulae system will result in the same calculation.

Preparing Solutions

It would have noted that most of the concentration terms were stated with reference to the volume of the solution. This final volume of the solution is not always equal to the sum of the volumes of the solute(s) and the solvent, although a solution is made up of solute(s) and a solvent. Therefore, in preparing a solution, the final volume of the solution must be measured. The chemical principle (partial molar properties of solution) explaining this concept states that the final volume of a solution or mixture is a function of (in other words, depends on) the moles of the components to be mixed. For example, when one mole of water (18 g and therefore 18 cm^3) is added to a large volume of a 70% ethanol solution, the volume increases by 16 cm^3 and not 18 cm^3. For some other solutions, the change in volume is higher than the added volume. Further discussion of this principle is beyond the scope of this book. We would consider the steps that may be followed when preparing a solution.

IN THE LABORATORY

The following steps can be considered in preparing a simple solution;

1. Calculate the amount (mass or volume) of solute needed to make-up a solution of a particular concentration and volume.

Always use the information on the label (molecular weight, % purity, etc.) of the container of the solute for the calculations.

2. Obtain a clean wash-glass, weighing boat or a weighing paper and carefully weigh out the calculated mass of solids. For a liquid, obtain a clean and dry measuring cylinder to measure the calculated volume.

3. Transfer completely, the solute into a clean beaker that can contain at least twice the final volume of the solution.

4. The solvent (the volume of which should be at most, $^2/_3$ of the final volume of the solution) must then be added to the solute and stirred to dissolve completely the solute.

   It should be noted that some solutes as they dissolve may get hot, warm (e.g NaOH) or cold (Tris-base). Also, some solute may not dissolve except at specific pH (EDTA dissolves at pH or higher).

5. For chemical/solutions stored at cold temperatures, they must be allowed to come to room temperature before proceeding with the preparation of the solution.

   For example, a solution sodium dodecyl sulfate would precipitate when added to a cold liquid substance/ solution.

6. If the pH of the solution is to be determined or maintained at a particular value, then this should be done at this stage, after which the solution must be transferred to a clean, dry volumetric flask of appropriate volume or a measuring cylinder if the solution might stain the volumetric flask (which makes it difficult to clean).

7. More solvent is then added to the mark of the final volume.

8. The solution must then be stored appropriately (considering their sensitivity to temperature and light *etc.*). The label of the chemical container and or the documents in the package may specify these conditions.

Let us use the following problems to illustrate some calculations associated with the preparation

Problem 3.1

Determine the final volume of a 0.5 M NaOH solution, when the solution is to be prepared using 10 g of NaOH. (The molar mass on the container is 40.0 g/mol).

Solution 3.1

The moles of NaOH in this solution $= \dfrac{\text{Mass}}{\text{Molar mass}} = \dfrac{10 \text{ g}}{40}$

$$n(\text{NaOH}) \text{ in this solution} = 0.25 \text{ mol}$$

$$\text{Concentration (molarity)} = \dfrac{\text{Number of moles}}{\text{Volume (dm}^3\text{)}}$$

$$\text{Volume (dm}^3\text{)} = \dfrac{\text{Number of moles}}{\text{Concentration (molarity)}} = \dfrac{0.25 \text{ mol}}{0.5 \text{ mol/dm}^3}$$

$$= \mathbf{0.5 \text{ dm}^3}$$

$$\text{Volume of solution to prepare} = \mathbf{500 \text{ cm}^3}$$

Problem 3.2

What volume of a 0.2 mol/dm^3 Na$_2$CO$_3$ solution will contain 26.5 g of the solute? (Assume the molar mass on the container as 106 g/mol).

Solution 3.2

Convert the concentration term mol/dm^3 to g/mol

$$\text{g/dm}^3 = \text{mol/dm}^3 \times \text{g/mol}$$

$$\therefore \text{Conc. (g/dm}^3\text{)} = 0.2 \text{ mol/dm}^3 \times 106 \text{ g/mol}$$

$$= 21.2 \text{ g/dm}^3$$

21.2 g is contained in 1000 cm^3

.: 26.5 g will be contained in $\dfrac{26.5 \text{ g}}{21.2 \text{ g}}$ x 1000 cm^3

$$= \textbf{1250 cm}^3$$

Problem 3.3

A solution of NaOH was prepared by adding water to 2.4 g of the solute until 250 cm^3 of solution was obtained. What is the concentration in (i) g/dm^3 (ii) mol/dm^3 (iii) % (w/v) of solution? (Assume the molar mass on the container as 40.0 g/mol).

Solution 3.3

(i) *From the problem*   250 cm^3 = 0.25 dm^3

0. 25 dm^3 contains 2.4 g of NaOH

.: 1 dm^3 will contain $\dfrac{1 \text{ dm}^3 \times 2.4 \text{ g}}{0.25 \text{ dm}^3}$ = 9.6 g

Concentration of NaOH = **9.6 g/mol**

(ii) Concentration in mol/dm^3 = $\dfrac{\text{Conc. in g/dm}^3}{\text{Molar mass g/mol}}$

Concentration in mol/dm^3 = $\dfrac{9.6 \text{ g/dm}^3}{40.0 \text{ g/mol}}$ = **0.24 mol/dm^3**

(iii) % concentration (w/v)

If 250 cm^3 contains 2.4 g of solute

Then 100 cm^3 will contain $\dfrac{100 \text{ cm}^3 \times 2.4 \text{ g}}{250 \text{ cm}^3}$ = **0.96% (w/v)**

17

Problem 3.4

A technician had to prepare 200 cm³ of a 0.450 M solution of hydrated sodium trioxocarbonate (IV) but had no information regarding the molar mass and the water of hydration (Na_2CO_3 XH_2O). However, when the technician dissolved 1.0 g of this compound to make a 5.0 cm³ solution, he determined the concentration to be 0.520 M. Calculate the

(i)   Molar mass

(ii)   Value of X

(iii)   Mass of this compound the technician needed to weigh in order to prepare the desired solution. (Na = 23; C = 12; O = 16; H = 1).

Solution 3.4

Form the problem

Mass = 4.00 g   Prepared volume = 200.0 cm³

Prepared conc. = 0.520 M

(i) $\text{Concentration} = \dfrac{\text{Mass (g)}}{\text{Molar mass (g/mol)} \times \text{Volume (dm}^3)}$

$0.52 \text{ mol/dm}^3 = \dfrac{4.00 \text{ g}}{Mr \times 0.0200 \text{ dm}^3}$

$Mr = \dfrac{4.00 \text{ g}}{0.520 \text{ mol/dm}^3 \times 0.0200 \text{ dm}^3}$

$Mr = \textbf{384.6 g/mol}$

(ii) $Mr = 2(23) + 12 + 3(16) + X (2 + 16) = 384.6$ g/mol

$$106 + X (18) = 384.6$$

$$X = 15.6$$

$$.: X = 16$$

(iii) Mass = Concentration × Molar mass (g/mol) × Volume (dm³)

$$= 0.450 \text{ mol/dm}^3 \times 384.6 \text{ g/mol} \times 0.200 \text{ dm}^3$$

$$\text{Mass} = \textbf{34.6 g}$$

Problem 3.5

A 50 mL solution of 6 μM p-nitrophenol in 0.02 M sodium hydroxide is to be prepared for an enzyme assay. How should this solution be prepared? The molecular weight of p-nitrophenol is 371.1 g/mole and that of sodium hydroxide is 40.0 g/mole.

Solution 3.5

We should first recognize that the solution of interest would be prepared in another solution as a solvent. Therefore, the first solution to prepare is the 0.02 M sodium hydroxide solution. The volume of this solution should be slightly more than 50 mL since it is to be used in making up the volume of the final solution. Therefore, we can decide to prepare 60 mL of the NaOH solution. To do this we need to calculate the mass of NaOH that would be needed.

Volume = 60.0 cm³   Conc. = 0.020 M

Molar mass of NaOH = 40.0 g/mol

(i) Mass (g) = Concentration × Molar mass × Volume

Mass (g) = 0.020 M × 40.0 g/mol × 0.060 dm³

Mass = **0.048 g**

Therefore, 0.048 g of NaOH would be dissolved in enough water and the volume made up to 60 mL.

The next step is to calculate the mass of the p-nitrophenol (pNP) needed to prepare the desired concentration.

19

Volume = 50.0 cm^3 Conc. = 60 x 10^{-6} M

Molar mass of pNP = 371.1 g

(ii) Mass = Concentration x Molar mass x Volume

Mass = 6 x 10^{-6} M x 371.1 g/mol x 0.05 dm^3

Mass = 0.00011133 g = **0.11133 mg**

This mass is obviously too small to weigh on most weighing balances in most laboratories. To be able to prepare solutions that present such difficulties, a more concentrated form of the solution is prepared after which it can be diluted to the desired concentration. This problem will be revisited after going through the section on dilutions (section 4).

Self Check

Two solid chemical were to be identified as NaOH and Na_2CO_3. A 2.12% (w/v) solution of one of these compounds, labelled A, was determined to be 0.2 M. 0.24 g of the second compound, labelled B, was used to prepare a 25.0 cm^3 solution whose concentration was determined to be 0.24 M. Help identify the compounds. (Na = 23; C = 12; O = 16; H = 1).

SECTION

4

DILUTION OF SOLUTIONS

Simple Dilution

The procedure presented in the previous section would not appropriate for the preparation of all types of solutions. Another method has been described for the preparation of these solutions. This is known as the **dilution** method. The dilution method is preferable:

1. When the desired solutions are commercially available as liquids and at higher concentration.

    A calculated volume of the commercial solution must be diluted to the desired lower concentration.

2. When the appropriate mass to weigh in order to prepare a solution of very low concentration is too small to weigh.

    This limitation is surmounted by preparing a smaller volume of a higher concentration of the solution (of a measurable mass) after which it is diluted to the desired lower concentration.

3. In cases where a higher concentration of the desired solution has already been prepared, it would be wasteful to prepare a new solution.

    A calculated volume of the solution of higher concentration can be diluted to the desired concentration.

4. Another important situation that would require the use of the dilution method is one in which a series of the same solution is to be prepared, such that they vary in concentration by a constant factor. It would be tedious to keep weighing the solute and dissolving it for each of the concentrations.

As stated earlier, the concentration of a solution is the amount of solute in either a unit volume or a particular volume of solution. Therefore, the addition of more solvent to a solution (when diluting the solution) will reduce the amount of solute per volume of the solution, thus reducing the concentration, but not the total amount of the solute in the solution. Therefore, the amount of the solute (moles) before and after the addition of the solvent will remain the same.

$$Moles\ (n) = Concentration \times Volume$$

Therefore, the concentration of the initial solution (C_1) multiplied by its volume (V_1) is equal to the concentration of the new solution (C_2) multiplied by its volume (V_2).

$$Then,\ C_1\ V_1 = C_2\ V_2$$

Knowing the initial concentration of the solution, the desired concentration and volume of the final solution, then the volume of the initial solution to be measured can be calculated using this equation:

$$V_1 = \frac{C_2 \times V_2}{C_1}$$

The solution to be diluted is referred to as the **Stock Solution:** a solution to be diluted in order to prepare a lower concentration of that solution. For commercially available liquid chemicals, the initial concentration has to be calculated using the information given on the bottle. The necessary information to look out for are;

1. The density or specific gravity (mass per volume)

2. The percentage purity or the percentage concentration (v/v)

3. The molecular weight of the chemical (mass per mol)

IN THE LABORATORY

The following steps can be considered in diluting a solution

1. After calculating and measuring the volume of the stock solution, V_1, it is transferred to a volumetric flask or a measuring cylinder.

2. More solvent is then added to make-up the final volume, V_2. As discussed in section three, the difference in volume is not to be added to the solution.

Based on this method of dilution, the convention of reporting the dilution process is to state the volume of the stock solution as a ratio of the total volume of the final dilute solution. For example, when 5 cm^3 of a stock NaOH solution is diluted by adding enough solvent to make a 50 cm^3 solution, the dilution process can be written as a **5 in 50** dilution, which is reported as a **1 in 10** dilution or 1/10 dilution. The advantage of this is that the dilution factor is the same as the ratio. Therefore, the concentration of the dilute solution is one tenth that of the stock solution.

Another convention of preparing and/or reporting a dilution process would require that the 5 cm^3 of stock NaOH is added to 45 cm^3 of solvent. This is summarized as **5 to 45** and reported as a **1 to 9** dilution. We note that the ratio here is not the same as the dilution factor and that the final volume may not be the same as expected (50 cm^3) when volumes are added (this concept was discussed earlier in section three). For instance, a 1 **in** 10 dilution of 1 M HCl may slightly be more concentrated than a 1 **to** 10 dilution of the same 1 M HCl solution. Note that the final volume in the former dilution is 10 mL while that of the latter is expected to be 11 mL.

Serial Dilution

To prepare a number of solutions of the same chemical whose concentrations vary progressively by the same dilution factor (1 in 10, 1 in 100, 1 in 1000 etc.), the same initial volume, V_1 is measured and transferred into the next tube of a series of separate tubes to contain the next lower concentration after the volume of the previous tube has been made to the mark. This procedure is known as **serial dilution** and is illustrated below for the preparation of a series of 10 cm^3 solutions with a concentration decreasing by a factor of 1/10.

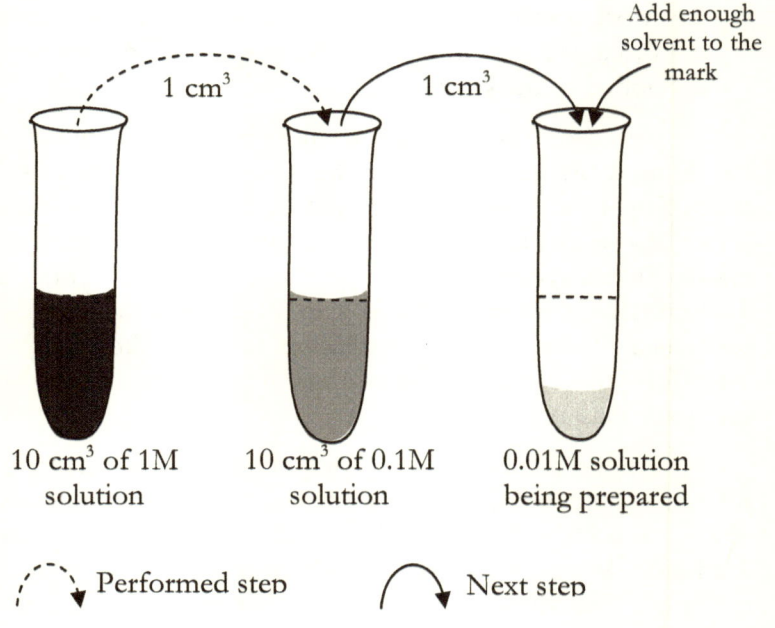

1 cm^3 1 cm^3 Add enough solvent to the mark

10 cm^3 of 1M solution

10 cm^3 of 0.1M solution

0.01M solution being prepared

Performed step   Next step

In considering the following examples, it would become clear how important it is to understand the dilution process.

Problem 4.1

What is the concentration of a 25 mL solution prepared by diluting 15 mL of 1.5 M NaOH with distilled water?

Solution 4.1

$C_1 = 1.5 \text{ M}$ $C_2 = ?$ $V_1 = 15 \text{ mL}$ $V_2 = 25 \text{ mL}$

$$C_2 = \frac{C_1 \times V_1}{V_2} \qquad C_2 = \frac{1.5 \text{ M} \times 15 \text{ mL}}{25 \text{ mL}}$$

$$C_2 = 0.90 \text{ M}$$

Problem 4.2

A student added 2 mL of a 0.5 M Tris-HCl solution to 100 mL of a 1.4 M solution of EDTA. Determine the concentrations of Tris-HCl and EDTA in this new solution if the student determines the final volume to be 102 mL.

Solution 4.2

In such determinations, one of the solutions (at a time) is considered as a solvent, contributing only to the increase in volume as would water while following the usual method of calculation involving dilutions. Also, it should be noted that the final volume (V_2) in such cases should be measured. However, the sum of the volume of the two solutions may be used as the final volume if the final volume is not measured after the dilution. We will begin by considering the EDTA solution as the solvent so as to determine the concentration of the Tris-HCl solution. Then we will do the reverse to determine the concentration of EDTA in the new solution.

a. $C_{1\,(Tris-HCl)} = 0.5\,M$ $C_{2\,(Tris-HCl)} = ?$ $V_{1\,(Tris-HCl)} = 2.0\,mL$

$V_2 = 102\,mL$

$$C_2 = \frac{C_1 \times V_1}{V_2} \qquad C_2 = \frac{0.5\,M \times 2.0\,\cancel{mL}}{102\,\cancel{mL}}$$

$$C_2 = 0.0098\,M \stackrel{\sim}{=} \mathbf{0.01\,M}$$

The concentration of Tris-HCl in the new solution is 0.01 M

b. $C_{1\,(EDTA)} = 1.4\,M$ $C_{2\,(EDTA)} = ?$ $V_{1\,(EDTA)} = 100\,mL$

$V_2 = 102\,mL$

$$C_2 = \frac{C_1 \times V_1}{V_2} \qquad C_2 = \frac{1.4\,M \times 100\,\cancel{mL}}{102\,\cancel{mL}}$$

$$C_2 = 1.3725 \stackrel{\sim}{=} \mathbf{1.4\,M}$$

The concentration of EDTA in the new solution is still 1.4 M

The concentration of EDTA in the new solution did not change significantly because the change in volume (2 mL) compared to its volume (100 mL) was relatively small. In such cases, the concentrations of the two solutions can be determined using the larger volume as the final volume. Confirm this by re-calculating the concentration of the Tris-HCL using a final volume (V_2) of 100 mL. However, when the volumes of the solutions involved are close in value to each other, such that the change in volume for each is significant, then the new concentration will be significantly different from the initial concentrations. The next problem shows this.

Problem 4.3

An enzyme-reaction solution was prepared by adding 3.0 mL of a 0.5 M solution of NaCl to 12.0 mL of buffer containing 5.0 mL of a 10% starch solution. Determine the final concentrations of NaCl and starch in the reaction solution. Assuming the final volume is the same as the sum of the individual volumes.

Solution 4.3

Let us begin with the determination of the final concentration of NaCl.

$$C_{1\,(NaCl)} = 0.5\ M \qquad C_{2\,(NaCl)} = ? \qquad V_{1\,(NaCl)} = 3.0\ mL$$

Final volume of solution $V_2 = 3\ mL + 12\ mL = 15\ mL$

$$C_2 = \frac{C_1 \times V_1}{V_2} \qquad C_2 = \frac{0.50\ M \times 3.0\ \cancel{mL}}{15\ \cancel{mL}}$$

$$C_2 \text{ of NaCl} = \mathbf{0.10\ M}$$

$C_{1 \text{ (starch)}} = 10\% \quad C_{2 \text{ (starch)}} = ? \qquad V_{1 \text{ (starch)}} = 5.0 \text{ mL}$

Final volume of solution $V_2 = 3 \text{ mL} + 12 \text{ mL} = 15 \text{ mL}$

$$C_2 = \frac{C_1 \times V_1}{V_2} = \frac{10\% \times 5.0 \text{ mL}}{15 \text{ mL}} \qquad C_{2 \text{ (starch)}} = 3.33\%$$

The concentration of starch in the new solution is 3.33%

Self Check

1. 25 cm³ of 2 M KOH solution is diluted to 500 cm². What is the molarity after dilution?

2. What volume of water should 250 cm³ of 2 M HCl be diluted with in order to obtain a 0.1 M solution of the acid.

3. Repeat the calculations for problem 3.7 above. Prepare a stock solution of a higher concentration.

4. A reaction solution is to contain the following solutions at the stated final concentration. The stock concentrations of these solutions are also given:

Solution	Working conc.	Stock conc.
Buffer	1X	10X
MgCl$_2$	2.0 mM	25 mM
dNTP mix	0.2 mM	2.5 mM
Primer 1	0.32 μM	20 μM
Primer 2	0.32 μM	20 μM
Enzyme	5 mg/μL	0.05 mg/μL

What volume of each of these solutions, will be required to prepare the reaction solution? Determine the extra volume required to make-up the final volume of the solution to 25 µL.

SECTION

5

BUFFER SOLUTIONS

Buffers

Buffers are solutions prepared to resist changes in pH when small amounts of acids and bases are added to them. A buffer, whatever the kind, consists of an acidic component and a basic component that would neutralize any added base or acid respectively. Two kinds of buffers can be prepared in a laboratory;

1. **"Related-Component" Buffer:** the most common type of buffer system prepared in most laboratories. This may either consist of a weak acid and its conjugate base available in the form of a simple salt or a weak base and its conjugate acid available in the form of a simple salt. These are also referred to as **Acid Buffers and Base Buffers** respectively. Examples of such buffers are the acetic acid/acetate buffer system and the ammonia/ammonium buffer system respectively.

2. **"Non-related Component" Buffer:** this type of buffer usually consists of two or more chemically different compounds that do not react with each other. While one acts as the acid, neutralizing any added base, the other acts as the base, neutralizing any added acid. An example is the Tris-EDTA buffer mostly used for nucleic acid extraction.

Planning a Buffer System

Before considering the step by step procedure of preparing a buffer, we have to consider a short review of the equation that relates the concentration of the components of a buffer and its pH as well as the planning that goes in preparing a buffer.

Suppose a weak acid HA and its conjugate base A⁻ are in solution, the proton equilibrium that is established is expressed as

$$HA + H_2O \overset{K_a}{\rightleftharpoons} H_3O^+ + A^-$$

The equilibrium constant, $K_a = \dfrac{[H_3O^+]_{eq}\,[A^-]_{eq}}{[HA]_{eq}}$

Taking -$\log_{10}$ of both sides, the equation becomes

$$-\log_{10} K_a = -\log_{10} [H_3O^+] - \log_{10} \dfrac{[A^-]_{eq}}{[HA]_{eq}}$$

The equation can be written as

$$pK_a = pH - \log_{10} \dfrac{[A^-]_{eq}}{[HA]_{eq}}$$

$$pH = pK_a + \log_{10} \dfrac{[A^-]_{eq}}{[HA]_{eq}}$$

$$pH = pK_a + \log_{10} \dfrac{[\text{Conjugate base }^-]_{eq}}{[\text{Acid}]_{eq}}$$

This expression, commonly known as the **Henderson-Hasselbalch** equation, is used in planning a buffer system. Based on the principles of common ion effect (which is outside the scope of this book), the initial molarity of both the base and acid components can be assumed to be equal to the equilibrium molarity of both the base and acid respectively. Therefore the pH can be expressed as

$$pH = pK_a + \log_{10} \dfrac{[\text{Base component}]}{[\text{Acid component}]}$$

Although the equation was expressed for an acid buffer, the same equation can be used for both acid and base buffers.

Suppose a buffer of a particular pH is to be prepared, the following suggested steps have been found useful to follow;

1. Determine the type of buffer to prepare, based on the pH that needs to be maintained. i.e. either acid or base buffers.

2. Decide on which weak acid or base to use. In taking this decision, a weak acid or base with a pK_a value close to the desired pH is preferred.

   a. *The rare alternative is to select a weak acid or base with a pK_a equal to the desired pH. This would mean that an equimolar solution of both the acidic and basic components of the buffer would be all that has to be prepared of which equal volume are mixed. The equation expressing this is shown below.*

   $$pH = pK_a + \log_{10}\frac{\chi}{\chi} \qquad\qquad pH = pK_a + 0$$

   $$pH = pK_a + \log_{10} 1 \qquad\qquad pH = pK_a$$

   b. *If the weak acid or base decided on is either a polyprotic acid or base (a weak acid or base that dissociates more once) care must be taken in choosing the components of the buffer. For example, the weak acid phosphoric acid dissociates as such*

$$H_3PO_4 \xrightleftharpoons{K_{a1}} H^+ + H_2PO_4^- \xrightleftharpoons{K_{a2}} H^+ + HPO_4^{2-} \xrightleftharpoons{K_{a3}} H^+ + PO_4^{3-}$$
$$pK_{a1} = 2.15 \qquad\qquad pK_{a2} = 7.20 \qquad\qquad pK_{a3} = 12.38$$

A phosphate buffer of pH close in value to its pK_{a1} (acid buffer) would have H_3PO_4 as its acidic component and $H_2PO_4^-$ as the basic component. A buffer of pH close to the pK_{a2} would require $H_2PO_4^-$ as the acid component and HPO_4^{2-} as the basic component. Similarly, a buffer of pH close to pK_{a3} would

31

have HPO_4^{2-} as the acidic component and PO_4^{3-} the basic component.

3. Except for the rare alternative stated above, the next step is to calculate the ratio of the molarity of the acidic and basic components of the buffer using the Henderson-Hasselbalch equation.

$$pH = pK_a + \log_{10}\frac{[Base]}{[Acid]}$$

$$pH - pK_a = \log_{10}\frac{[Base]}{[Acid]}$$

$$Antilog\,(pH - pK_a) = \frac{[Base]}{[Acid]}$$

4. The next step in planning is to decide on the molarity of the acidic and basic components. This should be done keeping in mind the amount (moles) of the strong acid or base that is expected to be introduced into the buffer. In line with this is the fact that a solution remains a buffer as far as the pH change is not greater than 1 pH unit. In other words, a solution is effective as a buffer in the **pH range** of $pK_a \pm 1$. Thus, in deciding the molarities, the moles of the acid component or base component should be chosen such that after the addition of the expected amount of strong base or acid respectively, the change in the ratio of the molarities would be within $1/10$ to $10/1$. This property of a buffer is what is referred to as the buffer capacity.

Buffer Capacity

Different definitions for **Buffer capacity** have been stated, one of which is stated as, the amount of acid or base that a buffer can neutralize before its pH changes significantly. The limitation of this particular definition is the use of the phrase "its pH changes significantly". A significant change in pH in one biological or chemical system may be by 0.1 pH units while in another system a change of 1.0 pH unit is what is considered as a significant change.

A definition that avoids this limitation is stated as the amount of strong acid or base that when added to 1 litre of buffer would result in a change of its pH by one unit. By this definition, the buffer capacity can be expressed as

$$\text{Buffer capacity, } \beta = \frac{dn}{dpH}$$

dn *is the change in moles of* H^+/OH *ions per litre*
dpH *is the change in pH*

In general, a maximum buffer capacity is achieved when the molarity of the components of the buffer are kept large and approximately equal. The molarity of the buffer, C_{buff} is the sum of the molarity of the acidic and basic components. For example, a 1 M acetate buffer may contain 0.5 M acetic acid and 0.5 M acetate ion or 0.7 M acetic acid and 0.3 M acetate ion.

These suggested steps in planning a buffer are not exclusive; therefore, we are encouraged to reason through any available set of steps using these as guide.

IN THE LABORATORY

Buffer Preparation: "Related Component Buffer"

The discussion so far is suggestive of only one of the methods of preparing a buffer. This we would refer to as the "**mixed-solute**" method that is, dissolving two different solutes in the same solvent. The alternative methods we would refer to as the "**mixed-solution**" and the "**single-solution**" methods. We will take a closer look at each of these methods in this section.

A. Mixed-Solute Method

After carefully planning for the buffer system;

1. The masses of the weak acid or base and the respective simple salt providing the conjugate base or acid needed for the preparation of the decided molarities are then calculated using the Henderson-Hasselbalch equation and weighted out separately as described for the preparation of solutions in section three.

2. These are then dissolved one after the other using a volume of solvent which is about 70% of the final volume of the buffer. This should be done noting all precautions stated earlier for preparing solutions.

3. The pH of the buffer is then checked with a standardised pH meter and adjusted with strong acid or base to the desired pH.

   a. *The needed standard buffers of pH 4.0, 7.0 and 9.0 are commercially available and should all be used for standardizing the pH meter before use. Although, the standard buffer of pH 7.0 can be used with either of the other two buffers depending on the pH that is to be measured, it is wrong to use only one of these buffers.*

   b. *Since pH is affected by temperature, both the standard buffers and the solution whose pH is to be determined should both be at the same temperature. Ideally, this temperature should be the temperature at which the solution is to be used. Therefore, a buffer picked from the fridge or cold room should be allowed to come to the appropriate temperature before measuring or adjusting its pH.*

   c. *In adjusting the pH of the prepared buffer, it is advisable to use concentrated acid or base, adding them a drop at a time with mixing. This would ensure that the final volume of the buffer is not exceeded by the time the desired pH is achieved. These concentrated acids and bases should be handled with all the necessary precaution since concentrated HCl may be fuming.*

4. The volume of the buffer is then made-up to the mark in a volumetric flask/graduated cylinder with solvent after which the pH is then rechecked.

B. *Mixed-Solution Method*

1. Calculate and weigh the masses of each of the components that would be needed to prepare a concentration that is equal to the concentration of the buffer. As an example, for a 0.5 M buffer, a 0.5 M solution of the acid and base components would be prepared separately. Prepare the same volume of each.

2. Gradually add one of the solutions to a small volume of the other while checking the pH.

3. When the desired pH is obtained, note the volume added and determine the volume ratio for the two volumes. This ratio is then scaled up to obtain the desired volume of the buffer. However, when the pH goes above the desired value, some of the other solution is used to bring it back to the desired pH.

The main advantage of this method is that it avoids the addition of more Na and Cl ions that will make the buffer not suitable for reactions that are sensitive to these ions. However, the major limitations of this method include;

1. More or less than necessary amount of the buffer may be prepared.

2. The moles of the individual components in buffer can only be calculated after the buffer has been prepared. This is because some amount of one of the components will be converted to the other. However, the concentration of the buffer remains the same as the initial concentration of the components.

C. *Single-Solution Method*

On some occasions, only a weak acid or the simple salt that provides its conjugate base may be the only reagent available in a laboratory. Under this condition a buffer can still be prepared by following the steps below;

1. The concentration of the buffer, which is the sum of the molarities of the weak acid and it conjugate base, is

decided/determined as in the previous method. This concentration is then used as the concentration of the available reagent and therefore used to calculate the mass of the available reagent that would be needed to prepare a solution of that concentration.

The mass of this component that would be needed to prepare this molarity is determined as is done for simple solutions, but not using the Henderson-Hasselbalch equation.

2. This mass is weighed and the solution is prepared as described for simple solutions earlier, without making the volume up to the final volume of the buffer.

3. The pH of the solution is checked with a standardized pH meter and a solution of a strong acid (HCl) or base (NaOH) is used to reduce or increase the pH respectively to the desired pH.

4. The volume of the solution can then be made-up to the mark in a volumetric flask

The disadvantage of this method is that an extra large amount of Cl^- or Na^+ is added to the buffer during the adjustment of the pH with the HCl and/or NaOH respectively. These ions in large amounts in solution may be either inhibitory or activating to some enzymes thereby rendering buffers prepared using this method less useful for buffering such enzyme reactions. There is also an increase in the ionic strength of the buffer which would make buffers prepared by this option, less useful for buffering reactions that are sensitive to the ionic strength of the reaction medium and for electrophoresis.

Another form of this third method of preparing a buffer is one in which the amount of acid (HCl) or base (NaOH) used for the adjustment of the pH is calculated. For this method the steps to follow include;

1. Using the principles of chemical reaction (discussed in the next section), the moles of the acid or base that would convert some of the available reagent to the required amount of the unavailable component in order to obtain a particular pH is calculated.

Note that this required amount of the unavailable component must first be calculated using the Henderson-Hasselbach equation as discussed for steps 3 and 4 of planning a buffer system on the pervious page.

2. The moles are used to calculate the mass or volume of standard solutions (solution of known concentration) of the acid or base needed.

3. Prepare the solution of the available reagent as in steps 1 and 2 of the above method.

4. Noting the final volume of the buffer, this mass of the acid or base is dissolved in a small volume of solvent after which it is added to the prepared solution of the available reagent. A small volume is used to avoid exceeding the final volume.

5. However, if the moles are converted to volume of a standard solution of the acid or base, then the two solutions are then mixed.
   Note that the concentration of the acid or base should be high enough so that only small volume (drops) of the acid or base would have to be added. This would ensure that the final volume of buffer would not become higher than desired.

6. The pH is then checked and the buffer made up to the final volume with solvent (distilled water).

D. Tables of Volumes

Using the methods stated so far, tables of volumes have been prepared to make it easier to prepare buffers. The use of such tables does not necessarily require an understanding of the steps in planning the preparation of a buffer. These tables report the volume and concentration of solutions of:

1. weak acids and salts that provide their conjugate bases,

2. weak bases and salts that provide their conjugate acids,

3. or the volume pairs of specific concentrations of weak bases and strong acids, or weak acids and strong bases,

that when mixed would result in a stated pH value of the desired buffer.

Examples

A table for the preparation of a sodium phosphate buffer is used as an example of a *mixed solution buffer*. Note that the disodium and monosodium salts of phosphoric acid are used in this table. This is in relation to the pH of the buffers and the pKa of the conversion of the monosodium to the disodium salt.

Table 5.1: Preparation of 100 cm^3 of a 0.1 M sodium phosphate buffer at 25°C.

Desired pH	Vol. 0.1 M Na$_2$HPO$_4$ (cm^3)	Vol. 0.1 M NaH$_2$PO$_4$ (cm^3)
5.80	7.9	92.1
6.00	12.0	88.0
6.20	17.8	82.2
6.40	25.5	74.5
6.60	35.2	64.8
7.00	57.7	42.3
7.20	68.4	31.6
7.40	77.4	22.6
7.60	84.5	15.5
7.80	89.6	10.4

For this type of table, the final volume and concentration of the buffer is not constant and can only be determined after the buffer has been prepared.

For a *single solution buffer*, the acetic acid buffer system is used as an example. The table 5.2 presents the volumes of 0.1 M solutions of the acidic component of the buffer (CH$_3$COOH) and the base (KOH) that would convert some of the acidic component to the basic component of the buffer.

Table 5.2: Preparation of acetic acid buffer systems at 25^0C.

Desired pH	Vol. of 0.1 M CH$_3$COOH (cm^3)	Vol. of 0.1 M KOH (cm^3)
4.20	43.03	9.93
4.40	30.87	9.95
4.60	23.14	9.97
4.80	18.38	9.98
5.00	15.34	9.99

IN THE LABORATORY

Buffer Preparation: "Non Related Component Buffer"

We will now look at the preparation of a non related component buffer. For this type of buffer, there would be no need for the use of the Henderson-Hasselbalch equation since the components are not related by pK$_a$ or pK$_b$.

The first consideration in planning the preparation of such buffers is to decide the working concentration of the acid and base components. The next step is to decide the concentration of a stock solution for each component that would be diluted to obtain the working concentration. After these, the following steps are used in preparing the buffer.

1. These stock solutions are prepared as described for simple solutions in section three.

2. Using the principles of dilution discussed in section four, the volumes of the stock solutions that would be needed to prepare

39

the desired working concentrations and final volume of the buffer, are calculated.

3. A volume of solvent that is about 30% of the final volume of the buffer (i.e. 30 mL of solvent, if a 100 mL of buffer is needed) is first measured.

   This is to avoid the direct mixing of the stock solutions that may be at high concentrations and as such may interact leading to the precipitation of one component or the other and in rare cases may result in a chemical reaction. The chemical principle explaining the effect of concentration on reactivity of a chemical compound is out of the scope of this book.

4. The calculated volumes of the stock solutions are measured and added to the volume of solvent one after the other while mixing the solution.

5. The pH is then adjusted with either a strong acid or base. After which the volume is made up to the final volume of the buffer.

Let us now look at how to handle some calculations involving the preparation of buffers.

Problem 5.1

What mass of $NaC_2H_3O_2$ must be dissolved in a volume of $HC_2H_3O_2$ to produce a 30.0 cm^3 solution with a pH of 5.09 and a final concentration of 0.25 M $HC_2H_3O_2$.

$$HC_2H_3O_2 + H_2O \rightleftharpoons H_3O^+ + C_2H_3O_2^- \qquad K_a = 1.8 \times 10^{-5}$$

Solution 5.1

$\underline{NaC_2H_3O_2}$

Mass = ?

$\underline{HC_2H_3O_2}$

Conc. = 0.25 M

$\underline{Buffer}$

Volume = 30.0 mL = 0.03 dm^3

pH = 5.09

$K_a = 1.8 \times 10^{-5}$.: $pK_a = -\log K_a = 4.75$

$$pH = pK_a + \log_{10} \frac{[\,Base\,]}{[\,Acid\,]} \qquad antilog\,(0.30) = \frac{[\,Base\,]}{0.25\ M}$$

$$pH - pK_a = \log_{10} \frac{[\,Base\,]}{[\,Acid\,]} \qquad 0.25\ M \times 1.995 = [\,Base\,]$$

$$antilog\,(5.09 - 4.75) = \frac{[\,Base\,]}{0.25\ M} \qquad [\,Base\,] = \mathbf{0.4987\ M}$$

$$C = \frac{m}{Mr \times V}$$

$$m = C\,(M) \times Mr\,(g/mol) \times V\,(dm^3)$$

$$Mr = 23 + 24 + 32 = 82$$

$$m = 0.4987\ M \times 82\ g/mol \times 0.03\ dm^3$$

$$mass = \mathbf{0.308\ g}$$

Problem 5.2

The pH of 25 mL of a buffer was changed from 3.9 to 2.9 when 0.5 mL of a 0.5 M of HCl was added. What is the buffer capacity of this buffer at a pH of 3.9?

Solution 5.2

Volume of HCl = 0.5 mL = 0.0005 dm³

Volume of buffer = 25 mL = 0.025 dm³

Since the change in pH was by one unit, the buffer capacity is equal to the equivalent mole of HCl added to 1000 mL of the buffer

$$n = C \times V$$

$$n = 0.5\ M \times 0.0005\ dm^3 = 0.00025\ moles$$

If 25 mL of this buffer required 0.00025 moles of HCl

Then 1000 mL will require $\dfrac{1000 \text{ mL} \times 0.00025 \text{ mole}}{25 \text{ mL}}$

The buffer's capacity at a pH of 3.9 = **0.01 moles of acid**

Self Check

1. Calculate the concentrations of the buffers and that of the components of the buffers presented in Tables 4.1 and 4.2.

2. Confirm the pH values of the buffers in each table for the given volumes. Note that there is the need to use the correct pK_a.

SECTION

6

DETERMINING CONCENTRATIONS OF SOLUTIONS

Although solutions would have been carefully prepared following the descriptions in the previous sections, there is the need to determine the exact concentration of a prepared solution. The concept of titration (volumetric analysis) can be used to determine the exact concentration of a prepared solution; this is referred to as standardizing the solutions. Also, spectrophotometery can be used for the determination of the concentration of coloured solutions and solutions that absorb UV light.

In volumetric analysis, the amount of substance in solution is estimated by determining what volume of that solution reacts with a known volume of another solution (standard solution). This is made possible by the fact that there is an equation governing every reaction involving chemical substances in solution. Therefore, prior to discussing the application of titration in the determination of the concentration of a solution, there is the need for a short review of some basic principles of chemical reaction and mole ratio; these two concepts are integral parts of volumetric analysis.

Chemical Equations

Chemical substances, when they react with each other, do so according to rules which depend on the nature of the reacting substances. This is commonly represented as a chemical equation. A **chemical equation** informs us of the molar proportion of the reacting chemical substances and the conditions and/or requirement(s) for the reaction.

For example,

$$4\,FeS_{2(s)} \;+\; 11\,O_{2(g)} \;\rightarrow\; 2\,Fe_2O_{3(s)} \;+\; 8\,SO_{2(g)}$$

$$\text{Reactants} \quad \rightarrow \quad \text{Products}$$

As written, the equation is read as, four molecules of solid iron sulphide reacts with eleven molecules of gaseous oxygen to form two molecules of solid iron oxide and 8 molecules of gaseous sulphur oxide.

Reactants are the chemicals that are consumed while the **products** are the chemicals that are formed in a chemical reaction. It is a good idea to always include the state of the chemicals involved in the reaction; as solid "(s)", liquid "(l)", gas "(g)" or aqueous "(aq)".

Balancing Chemical Equations

When solid magnesium reacts with aqueous hydrochloric acid, aqueous magnesium chloride and hydrogen gas are produced.

$$Mg_{(s)} \ + \ HCl_{(aq)} \rightarrow \ MgCl_{2(aq)} \ + \ H_{2(g)}$$

To write this correctly as a chemical equation, we will require knowledge of the molar proportion of all the chemicals represented in the equation. In order to be able to do this, the law of conservation of matter is used. This law states that the number and kind of atoms present as reactants at the beginning of a chemical reaction will be the same number and kind of atom present at the end of the reaction as products. Any chemical equation that has been written to show this equality is known as a **balanced equation**. In writing a balanced equation, most often, whole numbers are placed in front of a chemical formula (which changes the number of atoms of all the atoms in that chemical formula). The subscript numbers that are part of the chemical formulae are not altered although they are used as part of the amount of each atom present.

Let us consider the above equation,

Reactants	Products
1 Mg atom	1 Mg atom
1 H atom	2 H atoms
1 Cl atom	2 Cl atoms

There is no hard and fast rule for balancing a chemical equation; however, there are guidelines, one of which is known as **balancing by inspection**.

To balance the above equation by inspection, the following can be done. Placing 2 in front the reactant HCl increases the number of these atoms on that side of the equation to two each, which balances the number on the product side of the equation. Therefore, this equation can be correctly written as

$$Mg_{(s)} + 2\,HCl_{(aq)} \rightarrow MgCl_{2(aq)} + H_{2(g)}$$

Other equations such as those for redox reactions are balanced by a more complex method, which involves the balancing of charges and the inclusion of H_2O.

Mole Ratio

After understanding one of the basic rules governing a chemical reaction, the next useful step is to show how knowledge of the amount of substance and volume of one of the reactant can be used to determine or calculate the amount of substance of another reactant and/or of product formed. The relative proportion of atoms in a balanced equation is also a representation of the relative proportion of the mole of the reactants and products. For that matter, it can be stated that for the above reaction,

$$Mg_{(s)} + 2\,HCl_{(aq)} \rightarrow MgCl_{2(aq)} + H_{2(g)}$$
$$1\text{ mole} + 2\text{ moles} \rightarrow 1\text{ mole} + 1\text{ mole}$$

For every mole of Mg that reacts, two moles of HCl are consumed and one mole each of $MgCl_2$ and H_2 are produced.

Therefore the mole ratio of Mg to HCl is 1:2. This can be written as

$$\frac{\text{moles of Mg}}{\text{moles of HCL}} = \frac{1}{2}$$

Therefore, $1 \times$ moles of HCL $= 2 \times$ moles of Mg

The mole ratio of Mg to H_2 is 1:1

$$\frac{\text{moles of Mg}}{\text{moles of H}_2} = \frac{1}{1}$$

Therefore,   moles of H_2 = moles of Mg

It becomes clear that the mole ratio for any two of the chemicals in the equation can be obtained and used for the determination of the mole or concentration of the others. In using the mole ratios for calculating the moles of the product(s) formed or reactant(s) consumed, it is the mole ratio of the limiting reactant (reactant that is completely consumed) not the reactant in excess that is used. For example, the following reaction occurs in excess HNO_3 of known concentration, consuming a particular volume of a Na_2CO_3 solution of known concentration.

$$Na_2CO_{3\,(aq)} + 2HNO_{3\,(aq)} \rightarrow 2NaNO_{3\,(aq)} + CO_{2\,(g)} + H_2O$$

The mole ratio to use for calculating the mole of $NaNO_3$ or CO_2 formed should be the mole ratio of Na_2CO_3 to the respective compounds ($NaNO_3$ or CO_2).

$$\frac{\text{moles of Na}_2\text{CO}_3}{\text{moles of CO}_2} = \frac{1}{1}$$

Therefore,   1 X moles of CO_2 = 1 X moles of Na_2CO_3

$$\frac{\text{moles of Na}_2\text{CO}_3}{\text{moles of NaNO}_3} = \frac{1}{2}$$

Therefore,   1 X moles of $NaNO_3$ = 2 X moles of Na_2CO_3

Since HNO_3 was in excess, the amount of it that reacted would be less than the amount of it that was used for the reaction.

$$\frac{\text{moles of Na}_2\text{CO}_3}{\text{moles of HNO}_3} = \frac{1}{2}$$

Therefore, $1 \times$ moles of $\text{HNO}_3 = 2 \times$ moles of Na_2CO_3

We can now continue with the determination of the exact concentrations of solution using the volumetric analysis systems.

Simple Titration

Titration has been the most useful technique for the determination of the exact concentration of solutions (directly prepared or diluted) as well as the determination of the % composition and /or the % purity of solutions.

Apparatus

The common **Apparatus** used for simple titration include the following:

The **Burette**: the following precautions must be taken when using a burette

1. A burette must be thoroughly cleaned and dried to give good titration results.

2. Before filling up the burette, it must be rinsed thoroughly with the solution, running the solution through the jet.

3. There should be no air bubble in the jet after the burette is filled up. Should any air bubble be trapped in the jet, it must be removed by filling the burette beyond the zero mark and opening the tap fully to allow the solution to run through the jet thereby removing the air bubble.

4. No hot liquid should be poured into the burette – the instrument was calibrated at a particular temperature, usually 20^0C or 25^0C and therefore must be used at or close to such temperatures.

47

5. Readings should be taken with the eye at the same level as the surface of the liquid to avoid errors due to parallax. This precaution is applicable to all volume measuring apparatus.

The **Pipette:** in addition to the precautions for the burette;

1. When transferring a solution using a pipette, the tip of the pipette must be in contact with the wall of the receiving apparatus such that the solution being released will run along the wall of the receiving apparatus.

2. Also, no liquid left in the jet should be removed after transferring the measured volume into the flask.

The **Volumetric and Conical Flasks**: wash these thoroughly and dry before use.

IN THE LABORATORY

Presented below is a brief description of the steps involved in titrating a solution.

1. After measuring out the volumes and filling the burette, open the tap of the burette, such that the solution will flow drop-wise into the other solution in the conical flask which should contain an indicator for the reaction.

    While the solution is being added, swirl the conical flask with your right hand and control the tap of the burette with your left hand.

2. Stop adding the solution from the burette when the FIRST PERMANENT colour change is observed.

    This should be when just the entire initial colour is lost.

3. This whole process is repeated two more times as suggested by the table, for recording the reading, presented below.

    In recording the volumes of the solution used during the titration, the following format has been found useful and understandable by most scientists.

Example

Pipette Volume =

Indicator Used =

Colour change = from………..to………….

Burette reading (cm^3)	1st titration	2nd titration	3rd titration
Final volume	A	D	G
Initial volume	B	E	H
Volume used	A – B = C	D – E = F	G – H = I

$$\text{Average Burette reading (Titre Value)} = \frac{C+F+I}{3}$$

It is important to note that all the volumes are recorded to at least one decimal place (this depends on the burette being used). Also, it is NOT all the volumes recorded for the titrations that may be used in calculating the average titre value. It is only the volumes that are consistent (do not vary much from each other) that can be used in the calculation. Therefore, when there is an inconsistency, NONE OF VOLUME CAN AND MUST BE USED as titre value.

As a guide, the following examples should help in understanding how to set out to determine the concentration, volume, percentage purity etc of a solution using simple titration.

Problem 6.1

An alkaline solution containing 1.50 g of NaOH in 250 cm^3 of solution was titrated against 0.1 M HCl solution. What volume of the acid would be needed to react with 20 cm^3 of the alkaline solution?

Solution 6.1

The advisable first step is to summarize the information provided in the problem followed by converting all measurements into a common SI unit.

From this problem

	NaOH	HCl
Mass concentration (ρ)	$1.5 \text{ g}/250 \text{ cm}^3$	-
Molarity (M)	-	$0.1 \text{ mol}/dm^3$
Volume used (V)	20 cm^3	-
Moles in volume used (n)	?	?

<u>Converting mass concentration of NaOH to Molarity:</u>

From the information,

If 250 cm^3 of NaOH solution contains 1.5 g of NaOH

Then 1000 cm^3 of solution will contain $= \dfrac{1000 \text{ cm}^3}{250 \text{ cm}^3} \times 1.5 \text{ g}$

$$= \mathbf{6 \ g}$$

This means that 6 g of NaOH is contained in $1 \ dm^3$ of solution, therefore using the molar weight of NaOH; the mass concentration can be converted into the molarity.

Molar weight of NaOH = (23+16+1) g/mol = 40 g/mol

$$\text{Molarity of NaOH} = \frac{\rho \ (NaOH)}{Mr \ (NaOH)}$$

$$\text{Molarity of NaOH} = \frac{6 \ \text{g}/dm^3}{40 \ \text{g}/\text{mole}} = 0.15 \ \text{mol}/dm^3$$

The next step is to write down the balanced equation of the chemical reaction.

$$NaOH_{(aq)} \quad + \quad HCl_{(aq)} \quad \rightarrow \quad NaCl_{(aq)} \quad + \quad H_2O_{(l)}$$

Interpretation of the equation

One mole of NaOH reacts with one mole of HCl to produce one mole of NaCl and one mole of water

Mole ratio from the balanced equation is $\dfrac{1}{1} = \dfrac{n\ (HCl)}{n\ (NaOH)}$

Therefore, the number of moles of HCl that would react is equal to the moles of NaOH it would react with. Therefore, the mole of NaOH must then be determined.

The number of moles of NaOH that would react can be calculated using two different approaches;

Alternative A: Using a formulae

$$n(NaOH) = \text{Conc.} \times \text{Vol.}(dm^3)$$

$$= 0.15\ mol/\cancel{dm^3} \times (20/1000)\ \cancel{dm^3} = 0.003\ moles$$

Alternative B: Using proportions

The molarity of NaOH implies that

1000 cm^3 of this NaOH solution contains 0.15 moles

by proportion,

20 cm^3 of it would contain $\dfrac{20\ \cancel{cm^3}}{1000\ \cancel{cm^3}} \times 0.15\ moles$

$$= 0.003\ moles$$

Since the number of moles of NaOH is equal to the moles of HCl it would react with, then the (n) of **HCl = 0.003 mole.**

With both the molarity of the *HCl* solution (from the problem) and now the moles of it that would react, the volume of *HCl* solution that would react can be determined by any of the two alternative methods below;

Alternative A: Using a formulae

$$n(HCl) = Conc. \times Vol.(dm^3)$$

$$Vol.(dm^3) = \frac{n\ (HCl)}{Conc.}$$

$$Vol.(dm^3) = \frac{0.003\ \text{mol}}{0.1\ \text{mol}/dm^3}$$

$$Vol.(dm^3) = 0.03\ dm^3 = \textbf{30 cm}^3$$

Alternative B: Using proportions

The molarity of HCl implies that

0.1 moles of HCl is contained in 1000 cm^3 of this solution

By proportion,

0.003 moles should be contained in $\dfrac{0.003\ \text{moles}}{0.1\ \text{moles}} \times 1000\ cm^3$

$$= \textbf{30 cm}^3$$

Problem 6.2

After preparing a solution containing 0.5 moles of HCl, the volume was then made up to 1 dm^3 (1 L or 1000 cm^3) using distilled water. 30.0 cm^3 of this solution was found to neutralize exactly 25.0 cm^3 of a solution of Na_2CO_3. Calculate the concentration of the alkali (Na_2CO_3) in

(i) mol/ dm^3 (ii) g/ dm^3

Solution 6.2

Summary of the information provided in the problem

	HCl	Na_2CO_3
Molarity (M)	0.5 mol/dm³	-
Volume used (V)	30.0 cm³	25.0 cm³

Balanced equation of the chemical reaction is

$$Na_2CO_{3(aq)} + 2HCl_{(aq)} \rightarrow 2NaCl_{(aq)} + H_2O_{(l)} + CO_{2(g)}$$

Interpretation of equation

1 mole of Na_2CO_3 reacts with 2 moles HCl to produce 2 moles of NaCl, 1 mole of H_2O and 1 mole of CO_2.

Mole ratio from the balanced equation is $\dfrac{2}{1} = \dfrac{n\,(HCl)}{n\,(Na_2CO_3)}$

Therefore, $n\,(Na_2CO_3) = \dfrac{n\,(HCl)}{2}$

Alternative A: Using a formulae

$n\,(HCl) = $ Conc. × Vol (dm³)

$= 0.5$ mol/~~dm³~~ × $(30\,/1000\,)$ ~~dm³~~

$= \textbf{0.15 mol}$

Alternative B: Using proportions

The molarity of HCL implies that

1000 cm³ of this HCl solution contains 0.5 moles

by proportion,

30 cm³ of it would contain $(30$ ~~cm³~~ $/\,1000$ ~~cm³~~$) × 0.5$ moles

$= \textbf{0.15 mol}$

Calculating the moles of Na_2CO_3

By the mole ratio, $n(Na_2CO_3)$ = ½ × mole of HCl

The number of moles of Na_2CO_3 = ½ × 0.15 = **0.0075 mol**

Calculating the concentration of Na_2CO_3

The calculated mole of Na_2CO_3 is contained in the 25 cm^3 of solution and since concentration in mol/dm^3 is the number of moles in 1000 cm^3

Then by proportion, if 25.0 cm^3 contains 0.0075 mol

1000 cm^3 of it will contain $\dfrac{1000 \ cm^3}{25.0 \ cm^3}$ × 0.0075 mol

= **0.30 mol**

Therefore the molarity of Na_2CO_3 = **0.30 mol/dm^3**

Self Check:

1. Use the formulae approach to repeat the calculation of the concentration of Na_2CO_3.

2. After a 1 in 10 dilution of a stock solution of H_2SO_4, 28.4 cm^3 of this solution is required to neutralize completely 25.00 cm^3 of a 0.12 M NaOH solution. What is the molarity of both the stock and diluted solution of H_2SO_4?

3. What mass of NaOH would be needed to neutralize exactly 20.0 cm^3 of a solution containing 4 g H_2SO_4 per dm^3 of solution (Atomic masses: Na = 23, O = 16, H = 1, S = 32).

The next set of examples will assume that the titration has been performed and so a table of the titre volumes will be provided.

Problem 6.3

'A' is a solution of NaOH of concentration 2 g/dm³. 'B' is a solution of HCl. Titrate 'B', in the burette, against 25 cm³ of 'A'; using methyl orange as indicator. Calculate the average volume of the acid used. Using your results, calculate (i) the molarity of B (ii) the concentration of B in g/dm³. (Ar ; Na = 23, O = 16, Cl = 35.5).

Solution 6.3

Record the volumes of acid and base used

Pipette volume = 25 cm³
Indicator used = methyl orange

Burette reading (cm³)	Titration		
	First	Second	Third
Final	32.60	32.50	32.50
Initial	0.00	0.00	0.00
Volume used	32.60	32.50	32.50

$$\text{Average volume (titer value)} = \frac{32.60 + 32.50 + 32.50}{3}$$
$$= 32.55 \text{ cm}^3$$

Since all the three volumes determined are consistent, all three are used in calculating the average volume. From the problem, the titration was to determine the volume of the acid that would react with the base.

(i) *Summary of information*

Concentration of NaOH = 4 g/dm³

Molar mass of NaOH = (23 + 16 + 1) = 40 g/mol

Molarity of NaOH = mass concentration /molar mass

$= 4 \text{ g/dm}^3 \div 40 \text{ g/mol} = \textbf{0.1 mol/dm}^3$

55

The balanced equation for this reaction is

$$NaOH_{(aq)} + HCl_{(aq)} \rightarrow NaCl_{(aq)} + H_2O_{(l)}$$

Interpretation of equation

1 mole of NaOH reacts with 1 mole of HCl to produce 1 mole of NaCl and 1 mole of H_2O.

The mole ratio of HCl : NaCl = 1:1

Therefore, the moles of HCl = the moles of NaOH

Calculating moles of HCl

Using the formulae alternative

The mole of NaOH = Conc. x Vol.(dm³)

= 0.1 mol/~~dm³~~ x (25 /1000) ~~dm³~~

= **0.0025 moles**

Hence, the number of moles of **HCl = 0.0025 mol**

Calculating the concentration of HCl

Using the formulae alternative

$$n(HCl) = Conc. \times Vol.(dm^3)$$

$$Conc. = \frac{n(HCl)}{Vol.(dm^3)} \qquad Conc. = \frac{0.0025 \text{ moles}}{0.03255 \text{ dm}^3}$$

Conc. (HCl) = **0.07688 mol/dm³**

= **0.079 M**

Self Check:

Use the proportion approach to repeat these calculations.

(ii) Conc. (g/dm^3) of HCl = molarity (~~mol~~/dm^3) × molar mass (g/~~mol~~)

$$= 0.07688 \times (1+35.5)$$

$$= \textbf{2.92 g/dm}^3$$

Problem 6.4

A is a solution of tetraoxosulphate (VI) acid containing χ g of H$_2$SO$_4$ per 1 dm^3. B is NaOH solution containing 2 g of NaOH per 250 cm^3.

(a) Put A in the burette and titrate against 25 cm^3 portion of B using methyl orange as indicator. Record the volume of the pipette. Tabulate your readings and calculate the average volume of acid used.

(b) From your results

   (i) Calculate the conc. of A in mol/dm^3

   (ii) Calculate the value of χ

The equation of the reaction is

2 NaOH + H$_2$SO$_4$ → Na$_2$SO$_4$ + 2 H$_2$O

Solution 6.4

Information form the experiment
Pipette volume = 25 cm^3
Indicator used = methyl orange

	Titration		
Burette reading	First	Second	Third
Final (cm^3)	21.23	20.20	40.40
Initial (cm$_3$)	0.00	0.00	20.20
Volume used (cm^3)	21.23	20.20	20.20

$$\text{Average volume reading (titer value)} = \frac{20.20 + 20.20}{2}$$

$$= 20.20 \text{ cm}^3$$

Only two of the three determinations of the volumes of the acid have been used to calculate the titre value since only these are consistent

Summary of available information

	NaOH	H$_2$SO$_4$
Mass concentration	2 g/ 250 cm^3	-
Volume used (cm^3)	25.00	20.20

(i) *Calculating the concentration of NaOH*

If 250 cm^3 of NaOH solution contains 2 g of solute

Then 1000 cm^3 of NaOH solution will contain

(1000 cm^3 ÷ 250 cm^3) × 2 g = 8 g

.: Conc. of NaOH = **8 g/dm^3**

Molarity of NaOH = Conc. (g/dm^3) ÷ Molar mass (g/mol)

Molar mass of NaOH = 23 + 16 + 1 = 40 g/mol

Molarity of NaOH = 8 g/dm^3 ÷ 40 g/mol

$$= \textbf{0.2 mol/dm}^3$$

Interpretation of equation

$$2\,NaOH \ + \ H_2SO_4 \rightarrow Na_2SO_4 \ + \ 2\,H_2O$$

From the equation, 2 moles of NaOH react with 1 mole of H_2SO_4 producing 1 mole of Na_2SO_4 and 2 moles of H_2O.

Therefore, the mole ratio of

$$NaOH : H_2SO_4 = 2 : 1$$

Moles of H_2SO_4 = ½ moles of NaOH ....(1)

The methods used in the previous problems can be used to complete this calculation. Assuming that the previous methods are well understood, the method introduced here is a shortened procedure that gives the same result without converting the volume to dm^3.

Mole = Conc. x Volume

Therefore ....(1)   can be written as

Conc. x Volume of H_2SO_4 = ½ x Conc. x Volume of NaOH

Conc. x 20.20 cm^3 = ½ x 0.2 mol/dm^3 x 25.00 cm^3

$$Conc. = \frac{½ \times 0.2 \ mol/dm^3 \ \times \ 25.00 \ cm^3}{20.20 \ cm^3}$$

Conc. (Molarity) of H_2SO_4 = **0.12 mol/dm³**

Self Check:

1.   Use the proportion and formulae procedures as used for the previous problems to solve this problem.

2.   Use this new method to solve the previous problems.

(ii) Since χ g of H_2SO_4 per 1 dm³ is the concentration in g/dm³, the equation as shown below will be useful,

Conc. of H_2SO_4 (g/dm³) = Molarity (mol/dm³) × Molar mass (g/mol)

But the molar mass of H_2SO_4 = 2(1) + 32 + 4(16) = 98 g/mol

∴ Conc. of H_2SO_4 = 0.12 mol/dm³ × 98 g/mol

$$= 11.76 \text{ g/dm}^3$$

$$\therefore \chi = \textbf{11.76 g}$$

Problem 6.5

A is a solution of an acid H_2X containing 11.76 g per dm³. B is a 0.2 M sodium hydroxide solution. Put solution A in the burette and titrate against 25 cm³ (20 cm³) portion of B using phenolphthalein as indicator. Form the results calculate;

(i) The molarity of solution A

(ii) The molar mass of H_2X

(iii) The relative molar mass of the group X⁻.

The equation of the reaction is;

$$H_2X + 2\,NaOH \rightarrow Na_2X + 2\,H_2O$$

Solution 6.5

Information form the experiment

Indicator used was phenolphthalein. Pipette volume = 25 cm³, Concentration of solution A = 11.76 g/dm³, Concentration of solution B = 0.2 M.

	Titration		
Burette reading	First	Second	Third
Final (cm^3)	20.23	20.20	40.40
Initial (cm^3)	0.00	0.00	20.20
Volume used (cm^3)	20.19	20.21	20.20

$$\text{Average volume reading (titer value)} = \frac{20.19 + 20.20 + 20.21}{3}$$

$$= 20.20 \text{ cm}^3$$

It should be noted that this time all three values or determinations of the volume of acid have been used to calculate the titre value since all these are consistent

Summary of available information

	Solution A	Solution B (NaOH)
Conc.	11.76 (g/dm^3)	0.2 M
Volume used (cm^3)	20.20	25.00
Volume (dm^3)	0.0202	0.0250

(i) Interpretation of reaction equation

$$H_2X + 2\,NaOH \rightarrow Na_2X + 2\,H_2O$$

The equation of the reaction indicates that 1 mole of HX reacts with 2 moles of NaOH to produce 1 mole of Na$_2$X and 2 moles of water.

moles of NaOH = Concentration (M) × Volume (dm^3)

$$= 0.2 \text{ mol/dm}^3 \times 0.025 \text{ dm}^3$$

$$= 0.005 \text{ moles}$$

The mole ratio of $NaOH : H_2X = 2 : 1$

.: Number of moles of H_2X = ½ the moles of NaOH .... (1)

The number of moles of H_2X = ½ x 0.005 = **0.0025 moles**

$$Conc. = \frac{Moles}{Volume}$$

$$Conc. = \frac{0.0025}{0.0020}$$

Conc. (Molarity) of H_2X = **0.12 mol/dm³**

(ii) Molar mass of H_2X (g/mol)

$$Molar\ mass = \frac{Concentration\ (g/\cancel{dm^3})}{Molarity\ (mol/\cancel{dm^3})}$$

$$Molar\ mass = \frac{11.76\ g/dm^3}{0.12\ mol/dm^3}$$

Molar mass of H_2X = **98 g/moL**

(iii)   Let represent the relative molar mass of the atom X with $\cancel{Z}$

Then $2(1) + \cancel{Z} = 98$

$\cancel{Z} = 98 - 2 = 96$

The relative molar mass of X = 96

Self Check:

A is a solution containing 6.00 g/dm^3 of trioxonitrate (V) acid. B is a solution of a hydrated trioxocarbonate (IV), $Na_2CO_3 \cdot XH_2O$, containing 14.19 g of the trioxocarbonate per dm^3.

Put solution B into the burette and titrate against 25 cm^3 portion the volume of the pipette. Tabulate your reading and calculate the average volume used. From your result calculate

(i) The molarity of solution B

(ii) The value of X in $Na_2CO_3 \cdot XH_2O$

(iii) The percentage by mass of water of crystallization in $Na_2CO_3 \cdot XH_2O$

Spectrophotometry

Optical methods of determining the concentration of a compound in solution involve the measurement of the intensity of light absorbed or emitted when a solution of that compound is exposed to light of a particular wavelength. We will concern ourselves with two optical methods:

i.   **Colorimetry/Photometry**, involves the measurement of the intensity of visible light (380 nm – 760 nm) absorbed by a coloured solution.

ii. **UV spectrophotometry**, on the other hand, involves measuring the absorbance of ultraviolet (UV) light (200 nm – 400 nm) absorbed by a compound in solution.

The ability of a compound or a molecule to absorb light of a particular wavelength depends on its structure. Specifically, a molecule/compound would interact with a discrete amount of energy in a radiation (light) based on the arrangement and movement of electrons, within the different energy levels of its molecular structure. The part(s) of the molecule that provide the electrons involved in the absorption of specific wavelength of light are known as

63

chromophores. For example, conjugated aromatic rings are mostly responsible for the UV absorbance of molecules; these rings have delocalised electrons which can move from one energy level to other when the energy in the UV light is absorbed. Also, proteins are known to absorb maximally at 280 nm (UV light) because they contain aromatic amino acid residues namely tryptophan, tyrosin, and phenylalnine. As the concentration of a compound in solution increases, the amount of the chromphoric group also increases, as such, the intensity of light absorbed by the solution should increase correspondingly.

We will shortly take a look at how this relationship is used in determining the concentration of a compound that is naturally able to absorb visible or UV light. However, it is also possible to determine the concentration of a compound that do not naturally absorb visible or UV light. To achieve this, the following procedures may be used;

a. The compound of interest may be reacted with another compound or molecule to yield a *product that absorbs* within the visible or UV range of wavelengths.

b. The compound of interest may be reacted with excess of *another compound that will consume some of its chromphoric group that absorb* within the visible or UV range after the reaction. Then the reduction in the absorbance of the second compound can be used to determine the concentration of the compound of interest.

c. The pH of the solution containing the compound may be changed to *generate ionic forms* (particularly anionic forms) of the compound which may absorb within the visible or UV range.

Beer-Lambert's Law

The relationship between the intensity of light absorbed and the concentration of the absorbing compound in solution is described by a combination of the laws of Beer and Lambert.

When light of a particular wavelength is directed at a solution, not all the incident intensity emerges from the solution, some is absorbed by the solution.

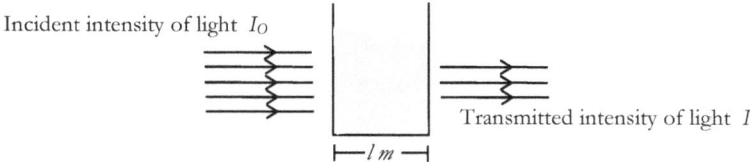

Incident intensity of light I_O

Transmitted intensity of light I

$\vdash l\, m \dashv$

The log to base ten of the ratio of the intensity of incident light (I_O) to the intensity of light that emerged from the solution (I) is a measure of the intensity of light that the solution absorbs. This is known as the **absorbance (A)**, which has no unit.

$$\text{Absorbance (A)} = \text{Log}_{10}\left(\frac{I_O}{I}\right) = \varepsilon cl$$

Where ε_λ = molar extinction coefficient of a compound at a wavelength, λ $(mol^{-1}m^{-2})$; c = concentration of the absorbing compound in solution (mole m^{-3}); l = the path length of light through the solution (m)

IN THE LABORATORY

We would have noticed that the Beer-Lambert's law was stated for the absorbance of a compound/molecule at a particular wavelength. This wavelength, known as the **wavelength of maximum absorption** (λ_{max}), is the wavelength of light at which the molecule absorbs the most; implying that the molecule may absorb at other wavelengths but not as much as it does at the λ_{max}.

Therefore, the first step in using spectrophotometry for the determination of the concentration of a compound/molecule would require knowledge of the λ_{max} of the compound/molecule. This can be obtained following the steps below:

1. The λ_{max} can be determined by exposing a fairly dilute solution of only that compound/molecule to a range of wavelength of light while measuring the corresponding absorbance.

2. A plot of the absorbance versus wavelength results in a curve known as the **Absorption spectrum**. The wavelength corresponding to the highest absorbance on this plot is the λ_{max}.

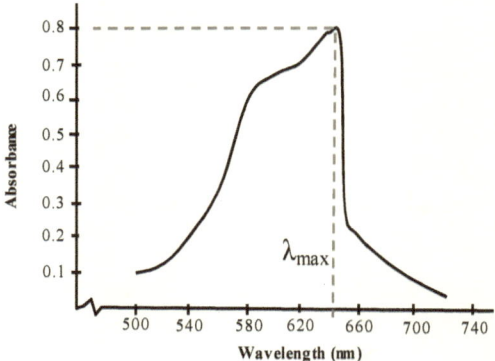

Fig. 7.1: Absorption spectrum of methylene blue.

3. The next step is to generate a calibration curve. To do this, different solutions (with a uniform increase in concentration) of the compound/molecule should be prepared.

Concentrations such as, 0.0 M, 0.2 M, 0.4 M, 0.6 M or any other concentration range may be used. However, it must be ensured that the solutions are fairly dilute. For coloured solution, this can be determined or achieved by using solutions that are fairly transparent.

The solution that does not contain the compound/molecule (concentration of 0.0 M) is used as a blank solution or a reference solution to zero the spectrophotometer. A **blank solution** is a solution that contains all the other components of a solution except the molecule whose absorbance (concentration) is being determined.

Therefore, every solution must have its peculiar blank solution. We should also note that the process of zeroing depends on the type of spectrophotometer being used.

The absorbance (at the determined λ_{max}) of each of the other solutions, starting from the most dilute to the most concentrated is then measured; re-zeroing the spectrophotometer with the blank solution before each measurement. This ensures that should there be a carry-over of some of the solution whose absorbance was previously measured to the next solution, the resultant change in concentration would be insignificant and so can be ignored.

A plot of the absorbance at the λ_{max} versus the corresponding concentration of the solutions is referred to as the **calibration** or **standard curve**. This plot is expected to be linear with an intercept at zero on both axes according to the Beer-Lambert's law as shown below;

$$A = (\varepsilon l)c + 0$$
$$y = (m)x + C$$

however, this plot is *not indefinitely linear*. Therefore, the best straight line should be drawn and the line should not be forced to intercept the both axes at zero. This is because: the accuracy of the spectrophotometer would not be uniform throughout the measurements, its precision is not 100% and that it is not equally sensitive at all concentration of the molecule.

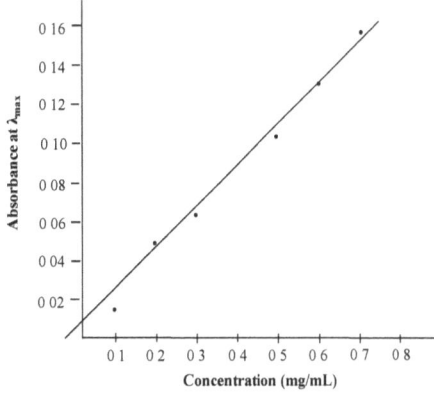

Fig. 7.2: A typical calibration curve.

4. The absorbance (at the determined λ_{max}) of the solution containing the compound/molecule whose concentration is to be determined can be measured using the appropriate blank solution. The absorbance is then interpolated on the calibration curve to determine its concentration.

Should this absorbance fall outside those used for the calibration curve, the straight line should not be extended to accommodate it, rather, the solution should be diluted appropriately. If the solution was taken through some process to generate its colour or to make it absorb UV light, then the starting solution should be diluted, after which the process is performed with the diluted solution. The absorbance can then be measured and the concentration of the dilute solution determined from the calibration curve. Using the dilution factor, as discussed in section four, the concentration of the original solution can then be calculated.

APPENDIX

Formulae

$$\text{Moles, n (mol)} = \frac{\text{Molarity (mol/dm)} \times \text{Volume (cm}^3 \text{ or mL)}}{1000}$$

$$\text{Moles, n (mol)} = \frac{\text{Mass, m (g)}}{\text{Molar mass, Mr (g/mol)}}$$

$$\text{Concentration, M (mol/ dm}^3) = \frac{\text{Mass, m (g)}}{\text{Molar mass, Mr (g/mol)} \times \text{Volume, V (dm}^3)}$$

$$\text{Concentration, M (mol/ dm}^3) = \frac{\text{Concentration, } \rho \text{ (g/dm}^3)}{\text{Molar mass, Mr (g/mol)}}$$

$$\text{\% concentration (w/v)} = \frac{\text{Mass of solute (g)}}{\text{Volume of solution (cm}^3)} \times 100$$

$$\text{\% concentration (v/v)} = \frac{\text{Volume of solute}}{\text{Volume of solution}} \times 100$$

$$\text{\% purity} = \frac{\text{Concentration of pure}}{\text{Concentration of impure}} \times 100$$

$$\text{\% impurity} = \frac{\text{Concentration of impure}}{\text{Concentration of pure}} \times 100$$

69

SI Units

SI units are the approved units by the general conference of weights and measures adopted by scientific laboratories worldwide. The basic SI units include the following;

Quantity	Symbols	Units	Abbreviation
Mass	m	kilogram	Kg
Time	T	second	s
Amount of substance	n	mole	mol
Temperature	Temp.	Kelvin	K
Length	L	Metre	m

Metric Multiples and Submultiples

The following prefixes have been used with the SI units.

Value	Prefix	Symbol	Value	Prefix	Symbol
10^{12}	tera	T	10^{-3}	milli	m
10^{9}	giga	G	10^{-6}	micro	μ
10^{6}	mega	M	10^{-9}	nano	n
10^{3}	kilo	K	10^{-12}	pico	p
10^{2}	hecto	h	10^{-15}	femto	f
10^{-1}	deci	d	10^{-16}	atto	a
10^{-2}	centi	c			

These prefixes are useful in avoiding the writing of too many zeros and larger or smaller digits. For instance, the number 0.00000075 mol is written as 0.75 μmol.

71

Bibliography

Davis, L.G., Kuehl, W.M., and Battey, J.F, *Basic Methods in Molecular Biology.* 2nd edition (New Jersey: Appleton and Lange, 1994).

Goldberg, D.E, *Fundamentals of Chemistry.* (Dubudge: Wm. C. Brown Publishers, 1994).

Greenbowe, T.J., Burke, K.A., and Pribyl, J, *A student Companion for Chemistry.* (New York: John Wiley and Sons. Inc., 1996).

Harley, P. J., and Prescott, M. L, *Laboratory Exercise in Microbiology.* 1st edition. (Dubudge: Wm. C. Brown Publishers, 1990).

Petrussi, R.H., and Harwood, W.S, *General Chemistry; Principles and Modern Application.* 6th edition (New Jersey: Prentice-Hall Inc, (1993).

Ragone, D.V, *Thermodynamics of Materials: the MIT Series in Material Science and Engineering.* Volume I (New York: John Wiley and Sons. Inc., 1995).

Wilson, K and Walker, J., *Principles and Techniques of Practical Biochemistry.* 4th edition (London: Cambridge University Press, 1992).

INDEX

About the Author

Adolf Kofi Awua was born in Accra, Ghana. He received his BSc. Honors degree in Biochemistry in 2001 from the University of Ghana, Legon. He spent four years as a Research and Teaching Assistant at the Biochemistry Department, University of Ghana, where he was involved with the study of microbial biodiversity using molecular techniques.

He earned his M.Phil. in Biochemistry from the University of Ghana, Legon in 2008. During this period, he studied the prevalence Human Papillomavirus in cervical cancers, using molecular genotyping techniques and also served as a Laboratory Demonstrator to the final year Molecular Biology practical course for two academic years.

He joined the Cellular and Clinical Research Centre, Radiological and Medical Sciences Research Institute, GAEC, in 2008 and is currently a Research Scientist. He stands with a rich experience in the laboratory and a good working understanding of student difficulties with solution preparation, as he has assisted with a good number of undergraduate dissertations during his working period at the University of Ghana, Legon.

www.ingramcontent.com/pod-product-compliance
Lightning Source LLC
Chambersburg PA
CBHW022121170526
45157CB00004B/1707